SPUNKY
SCIENCE

Copyright © 2022 Spunky Science

CAN:

✓ Make copies for your students for educational use

✓ Print content in different forms such as a booklet

✓ Print in various sizes to fit your needs

✓ Post content on a school-based platform for student use or reference

CAN'T:

✗ Distribute digital or copies to others without an additional purchase

✗ Remove Spunky Science logo or copyright

✗ Resell or redistribute in any way other than originally intended by Spunky Science

THIS BELONGS TO

_____

Spunky Science ©

SAFETY

Hair tie for long hair

Splash goggles

Lab coat or apron

Closed-toe shoes

disposable gloves

eye wash station

fire extinguisher

FIRE BLANKET

LAB SAFETY RULES:

- Do not eat or drink in the lab
- Wear appropriate PPE at all times
- Stay on task
- Follow lab instructions
- Dress appropriately
- Measure accurately
- Be careful when handling hot glassware
- Keep a clean work station
- Notify the teacher if glass breaks or equivalent

Sign here

print here

SCIENCE TOOLS

ruler and meter stick measure length

graduated cylinder

measuring liquids or displacement of small objects

Reading the meniscus

graduated cylinders come in all sizes

Beaker-measuring larger liquid volumes

300mL
250mL
200mL
150mL
100mL
50mL

Hot Plate: heating without a flame

Spunky Science ©

tens place →

| 0 | 10 | 20 | 30 | 40 | 50 | 60 | 70 | 80 | 90 | 100 |

hundreds place →

| 0 | 100 | 200 | 300 | 400 | 500 |

ones and decimal place →

| 0 | 1 | 2 | 3 | 4 | 5 | 6 | 7 | 8 | 9 | 10 |

Triple beam balance (measuring the mass of small objects

SCIENTIFIC REASONING

A descriptive statement that answers the question- what side are you on?

CLAIM

EVIDENCE

Two or three facts or Scientific data that directly supports your claim- why should we believe you?

REASONING

A full explanation as to how your evidence supports the claim that you have made.

INTERPRETING DATA

CHARTS

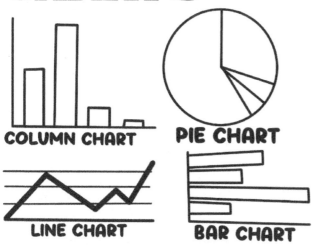

COLUMN CHART   **PIE CHART**

LINE CHART   **BAR CHART**

Charts are tools that organize information in order to easily see trends, make comparisons, and see the meaning behind the numbers.

TABLES

Element Abundance	%
Oxygen	47
Silicon	28
Aluminum	8
Iron	5
Calcium	4
Sodium	3
Potassium	2.6
Mgnsium	2

A table is a set of data that is organized using vertical columns and horizontal rows.

GRAPHS

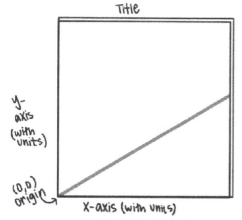

Title

y-axis (with units)

(0,0) origin

X-axis (with units)

Graphs take information, organize it, and present it as a "picture" of your data.

MAPS

A map is a representation of Earth's features drawn on a flat surface. Symbols and colors represent important features on a map.

ICE CREAM CONSUMPTION IS CAUSING SHARK ATTACKS

CORRELATION DOES NOT IMPLY CAUSATION

Spunky Science ©

BIOTIC AND ABIOTIC

Bio-LIFE
ORGANISMS COMPETE FOR BIOTIC THINGS FOR SURVIVAL SUCH AS PLANTS AND ANIMALS FOR FOOD.

A-NOT
ORGANISMS ALSO COMPETE FOR MANY NON LIVING THINGS FOR SURVIVAL SUCH AS TEMPERATURE, LIGHT, AIR, WATER, AND SOIL.

Spunky Science©

Label all the biotic and abiotic things that you see in this environment that help this Alligator survive. How many did you find?

STRUCTURES

Plants and animals have internal & external structures that supports SURVIVAL, GROWTH, BEHAVIOR, and REPRODUCTION.

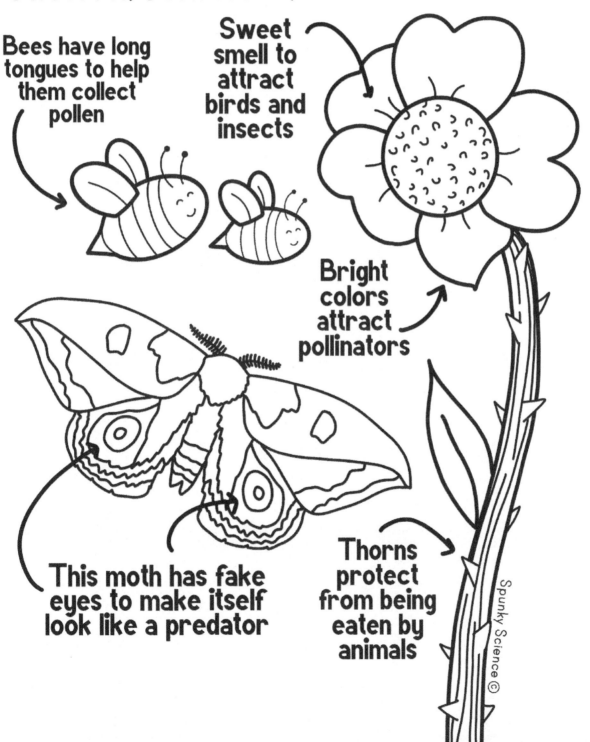

Bees have long tongues to help them collect pollen

Sweet smell to attract birds and insects

Bright colors attract pollinators

This moth has fake eyes to make itself look like a predator

Thorns protect from being eaten by animals

Spunky Science ©

The Love Dance of the BIRD of Paradise

FRONT VIEW

SIDE VIEW

These complex courtship performances can be broken down into a series of smaller, individual movements. These building blocks of motion are combined to form a single choreographed piece.

They hop, swing, strut, shake and buzz, and even transform their bodies into strange shapes to woo potential mates!

Over time, genes associated with these aesthetically pleasing features are passed down and the attributes become more prominant within the species.

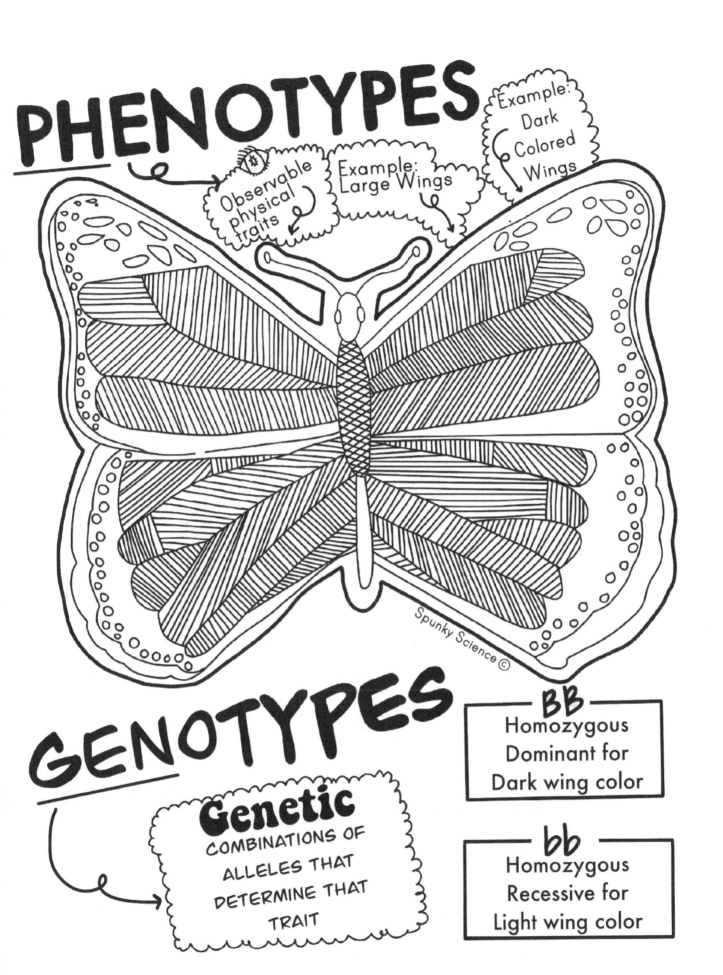

TRAITS

PHYSICAL CHARACTERISTICS

INHERITED
Genetically passed down from parent to offspring

Beak shape

Wingspan

Dimples

Fur

Petal Color and Shape

ACQUIRED
Influenced by experience or the environment

Scar

Taste in music

Accent

Weight

Spunky Science ©

3 TYPES OF CELLS

ANIMAL CELL

A eukaryotic cell that lacks a cell wall and has a true, membrane-bound nucleus.

PLANT CELL

A eukaryotic cell that has a cell wall, as well as other specialized cells that allow it to perform photosynthesis.

BACTERIA CELL

A prokaryotic, single-celled organism that lacks a nucleus and most other organelles.

Spunky Science ©

GENETICS

AND MENDELS PEA PLANTS

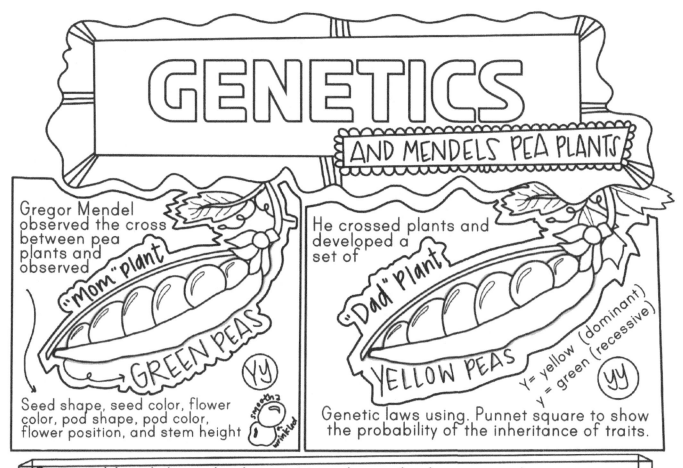

Gregor Mendel observed the cross between pea plants and observed

"Mom" plant

GREEN PEAS

Yy

Seed shape, seed color, flower color, pod shape, pod color, flower position, and stem height

smooth

wrinkled

He crossed plants and developed a set of

"Dad" plant

YELLOW PEAS

Y= yellow (dominant)
y = green (recessive)

yy

Genetic laws using. Punnet square to show the probability of the inheritance of traits.

Gregor Mendel studied genetics through observing the characteristics of peas through the use of a punnet square.

PUNNET SQUARE

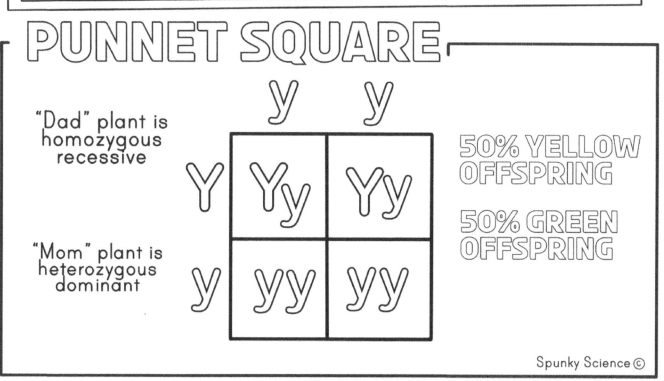

"Dad" plant is homozygous recessive

"Mom" plant is heterozygous dominant

	y	y
Y	Yy	Yy
y	yy	yy

50% YELLOW OFFSPRING

50% GREEN OFFSPRING

LEAFY SEADRAGON
Scientific name: Phycodurus eques

Leafy seadragon are not strong swimmers, so they rely heavily on their camouflage.

up to 30cm long

Eat plankton and small crustaceans

Fish Anatomy

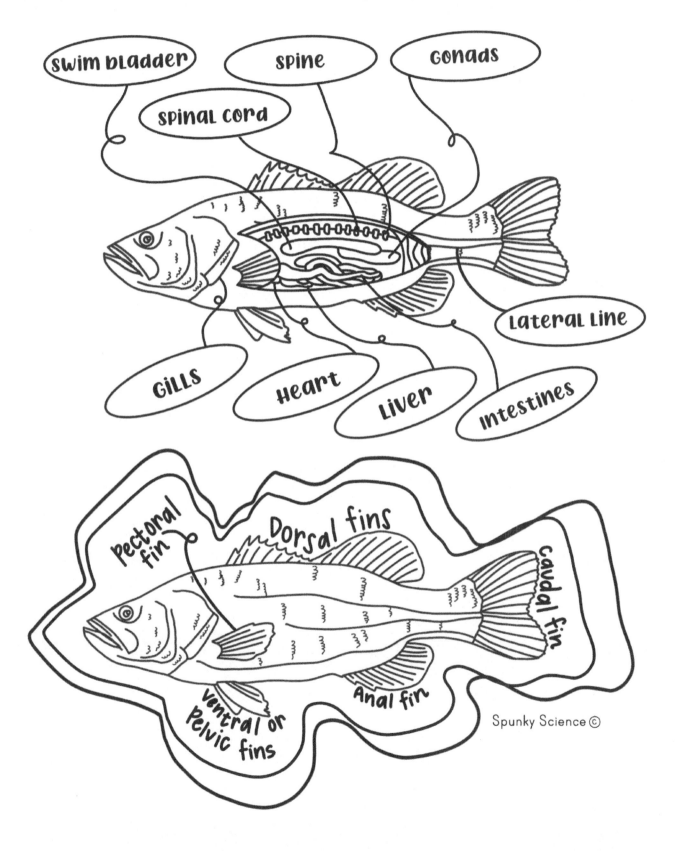

swim bladder

spinal cord

spine

gonads

lateral line

gills

heart

liver

intestines

Pectoral fin

Dorsal fins

caudal fin

ventral or pelvic fins

Anal fin

Spunky Science ©

AXOLOTL

Scientific name: Ambystoma mexicanum

Live up to
10 years

External gills

Coastal grooves
are used in
thermoregulation
and help to keep
the skin moist.

Diet: mollusks, worms,
insect larvae, crustaceans,
and some fish.

Axolotls have the ability
to completely
regenerate an entire
limb when lost!

Dorsal fin

STONEFISH
SCIENTIFIC NAME: SYNANCEIA VERRUCOSA

STONEFISH ARE THE MOST VENOMOUS FISH IN THE WORLD

USUALLY FOUND IN ROCKY REEFS NEAR THE COAST

EXCELLENT AT CAMOUFLAGE AND VERY FAST SWIMMERS

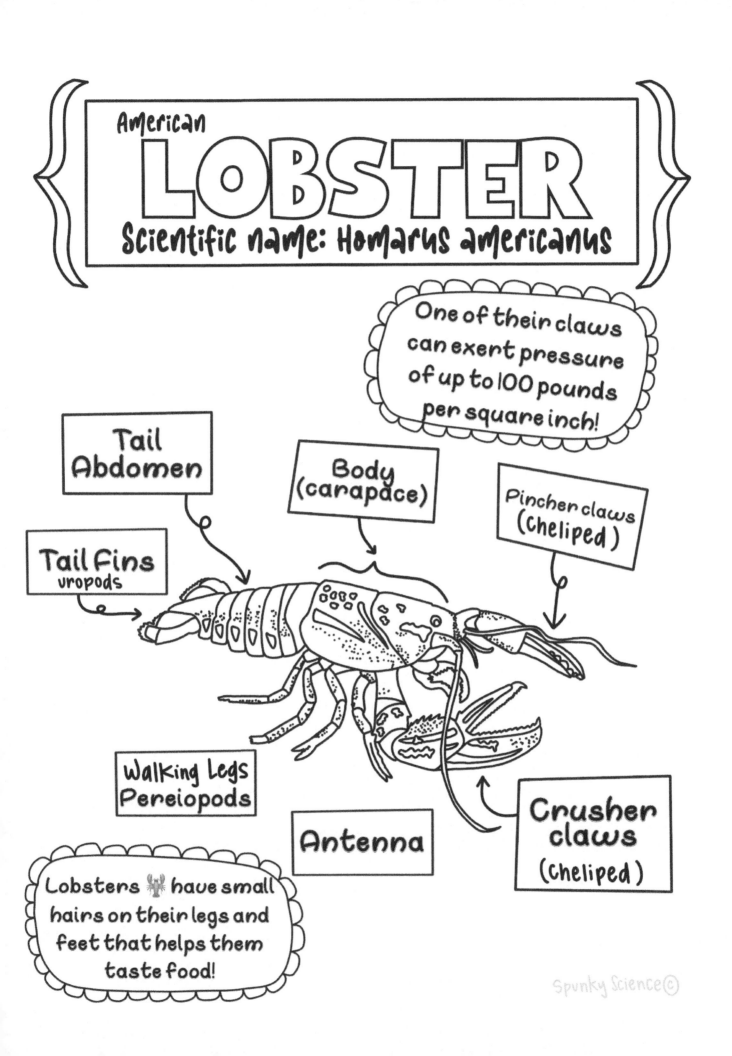

American
LOBSTER
Scientific name: Homarus americanus

One of their claws can exert pressure of up to 100 pounds per square inch!

Tail
Abdomen

Body
(carapace)

Pincher claws
(Cheliped)

Tail Fins
uropods

Walking Legs
Pereiopods

Antenna

Crusher
claws
(Cheliped)

Lobsters 🦞 have small hairs on their legs and feet that helps them taste food!

TURGOR PRESSURE

HIGH TURGOR PRESSURE

Vacuole is large and full of water in each cell

Plant Cell

Leaves have structure

LOW TURGOR PRESSURE

Plant Cell

Vacuole nearly empty

Cell wall sagging

Wilted leaves and stem

XYLEM

PHLOEM

TROPISM

Spunky Science ©

POSITIVE GEOTROPISM

Roots always grow with the direction of the gravitational pull

NEGATIVE GEOTROPISM

Leaves and stems always grow against gravity

GEOTROPISM (GRAVITROPISM)

How a plant responds to gravity

PHOTOTROPISM

When a plant responds to light

A flower grows towards the sun

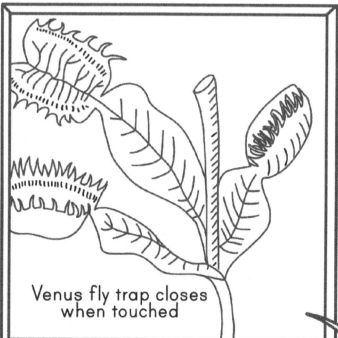

Venus fly trap closes when touched

THIGMOTROPISM

When a plant responds to touch

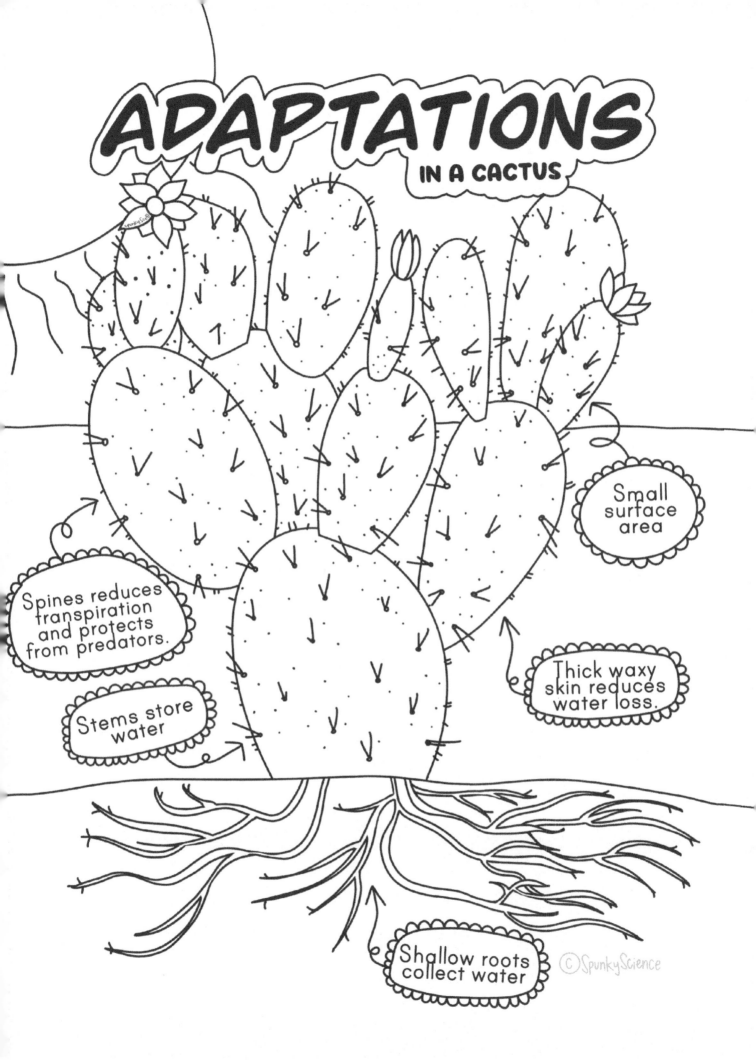

ADAPTATIONS

What is that? A change or process of change by which an organism becomes better suited to fit its environment.

Discovered By... Two French scientists, named Charles Darwin and Alfred Wallace, developed a theory that states that organisms have traits that are passed down that allow organisms to survive in their environments.

Can be BEHAVIORAL OR ANATOMICAL

The way a plant or animal acts

How the plant or animal is structured

Such as

GIRAFFES with long necks are tall enough to eat and are more likely to survive.

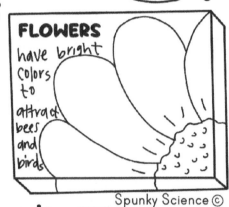

FLOWERS have bright colors to attract bees and birds

Spunky Science ©

Fur keeps BEARS warm.

PUFFER FISH take in water to look big and scare away predators.

CACTI have thorns so predators won't want to eat them.

ARTIFICIAL REEFS

An artificial reef is a man made structure that mimics the benefits of a natural reef.

Reefs are responsible for attracting fish, filtering water, increasing tourism, preventing erosion, and keep ecosystems healthy.

Most common artificial reefs are sunken ships, however some are made from concrete blocks.

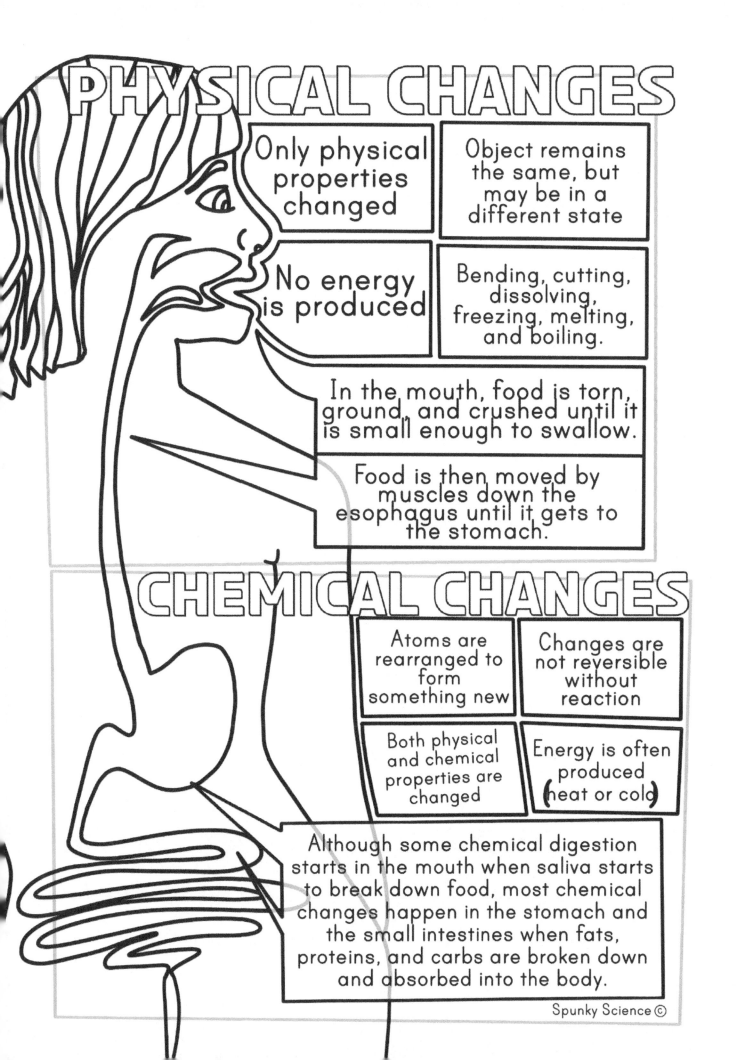

PHYSICAL CHANGES

Only physical properties changed

Object remains the same, but may be in a different state

No energy is produced

Bending, cutting, dissolving, freezing, melting, and boiling.

In the mouth, food is torn, ground, and crushed until it is small enough to swallow.

Food is then moved by muscles down the esophagus until it gets to the stomach.

CHEMICAL CHANGES

Atoms are rearranged to form something new

Changes are not reversible without reaction

Both physical and chemical properties are changed

Energy is often produced (heat or cold)

Although some chemical digestion starts in the mouth when saliva starts to break down food, most chemical changes happen in the stomach and the small intestines when fats, proteins, and carbs are broken down and absorbed into the body.

Spunky Science ©

ENERGY IN ECOSYSTEMS
DRAW ORGANISMS THAT BELONG IN EACH CATEGORY!

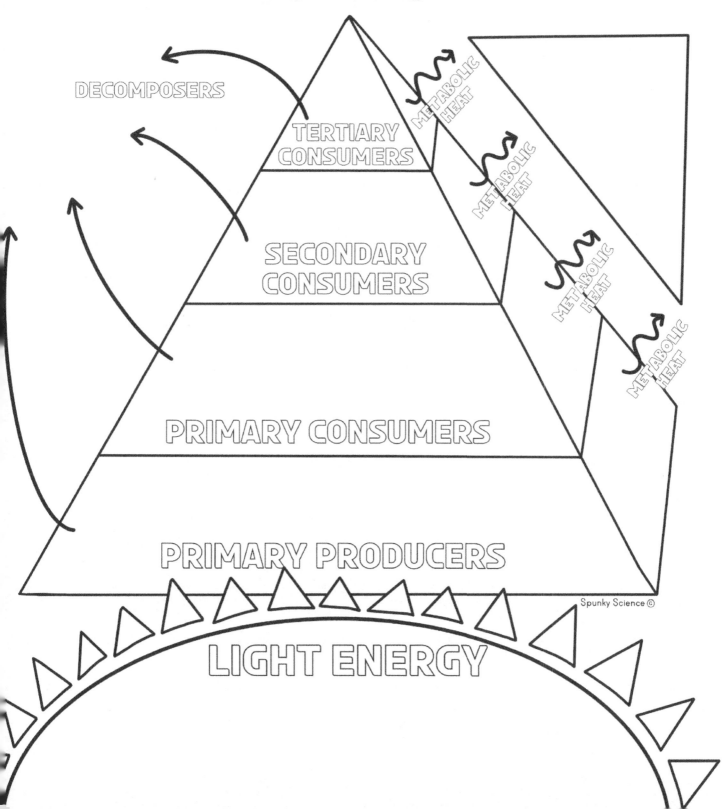

DECOMPOSERS

TERTIARY CONSUMERS

METABOLIC HEAT

METABOLIC HEAT

METABOLIC HEAT

METABOLIC HEAT

SECONDARY CONSUMERS

PRIMARY CONSUMERS

PRIMARY PRODUCERS

Spunky Science ©

LIGHT ENERGY

DECOMPOSERS

Mushrooms/Fungi

Bacteria

Earth worms

Decomposes are organisms that break down dead organic material.

Spunky Science ©

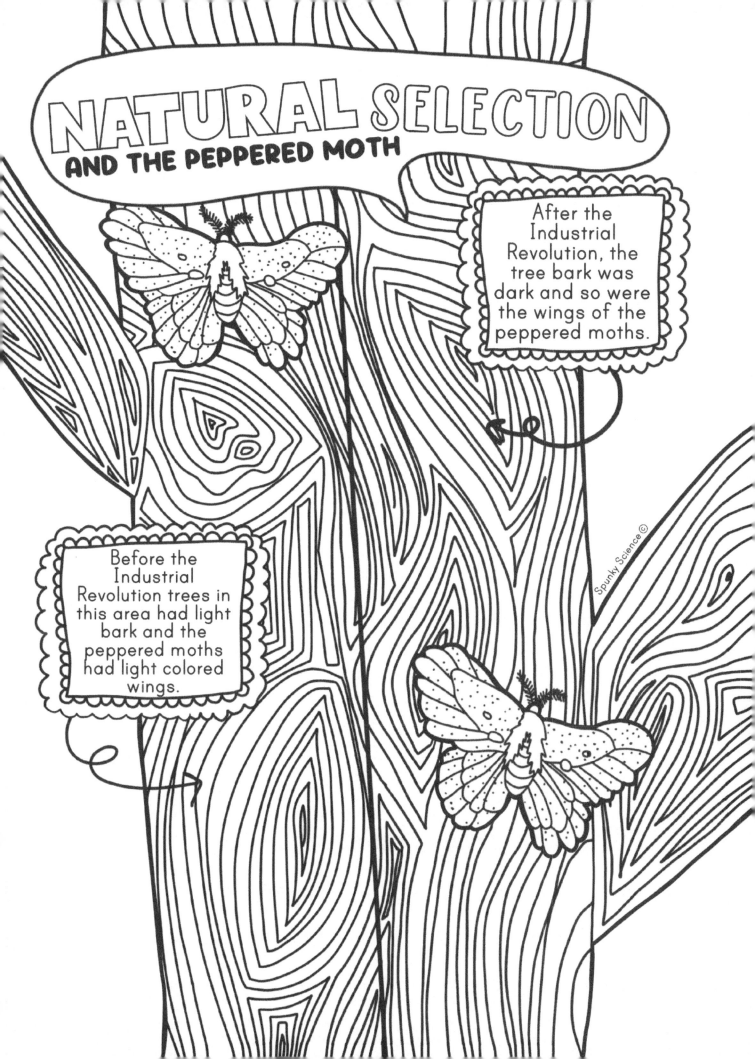

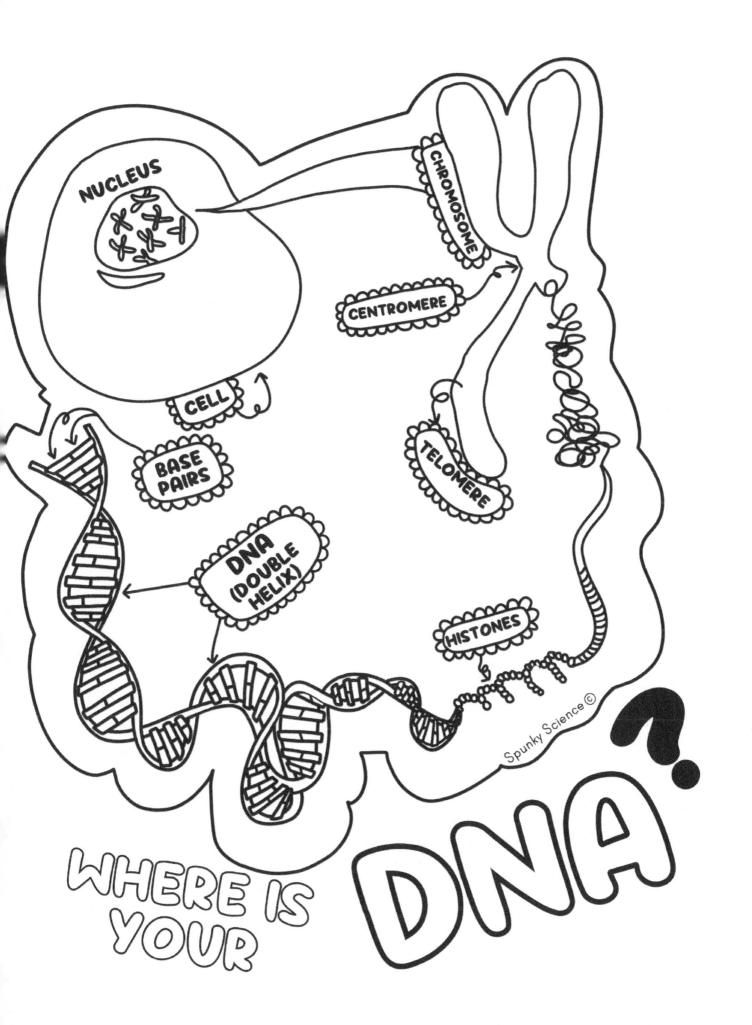

Levels of biological organization

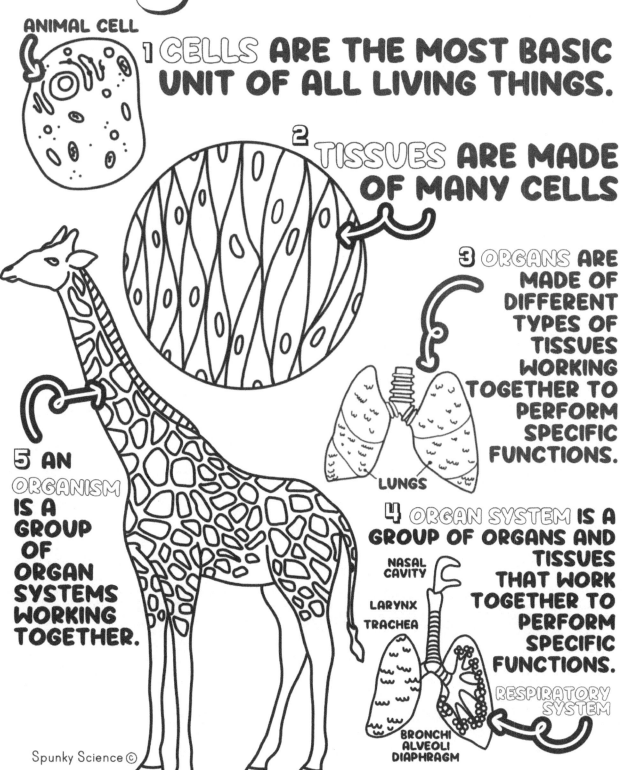

ANIMAL CELL

1 CELLS ARE THE MOST BASIC UNIT OF ALL LIVING THINGS.

2 TISSUES ARE MADE OF MANY CELLS

3 ORGANS ARE MADE OF DIFFERENT TYPES OF TISSUES WORKING TOGETHER TO PERFORM SPECIFIC FUNCTIONS.

LUNGS

5 AN ORGANISM IS A GROUP OF ORGAN SYSTEMS WORKING TOGETHER.

4 ORGAN SYSTEM IS A GROUP OF ORGANS AND TISSUES THAT WORK TOGETHER TO PERFORM SPECIFIC FUNCTIONS.

NASAL CAVITY

LARYNX

TRACHEA

RESPIRATORY SYSTEM

BRONCHI
ALVEOLI
DIAPHRAGM

THE BRAIN AND THE NERVOUS SYSTEM

Spunky Science ©

Connects with the central nervous system to control the body.

CNS= Brain & spinal cord

Blood Supply

Blood is carried to the brain by two paired arteries

Cerebrum
- Speech
- Reasoning
- Emotion

Cerebellum
- Balance
- Posture
- Coordinate Muscles

Brain Stem
- Relay System for automatic function

PNS

Peripheral Nervous Sstem
- All nerves outside the brain and spine

Cells of the Brain

The brain is made up of two types of cells - nerve cells and glia cells.

Nerve Cells (Neurons)

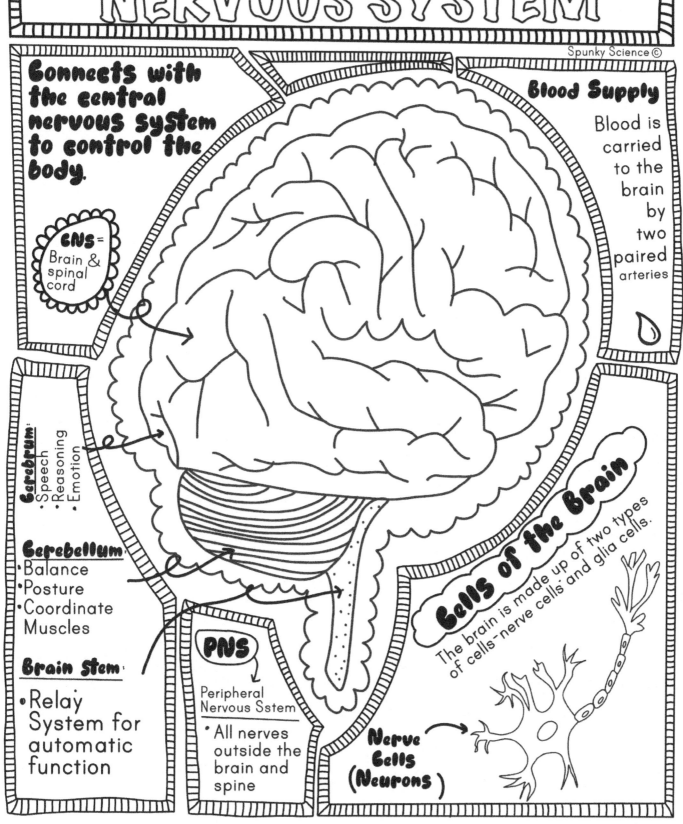

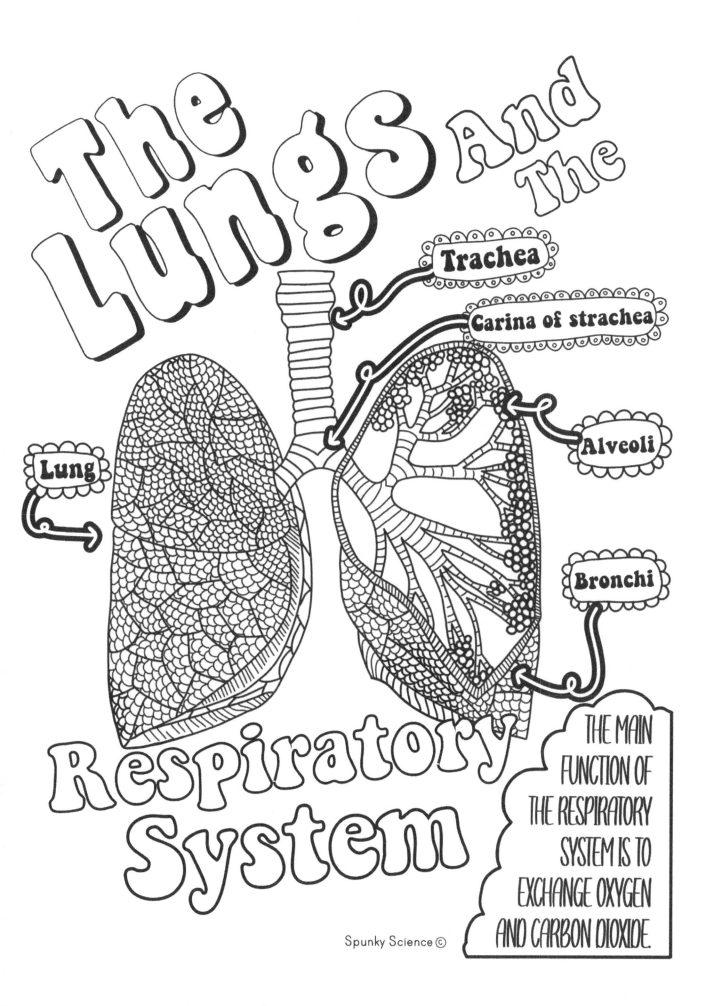

The Lungs And The Respiratory System

Trachea

Carina of strachea

Alveoli

Lung

Bronchi

THE MAIN FUNCTION OF THE RESPIRATORY SYSTEM IS TO EXCHANGE OXYGEN AND CARBON DIOXIDE.

Spunky Science ©

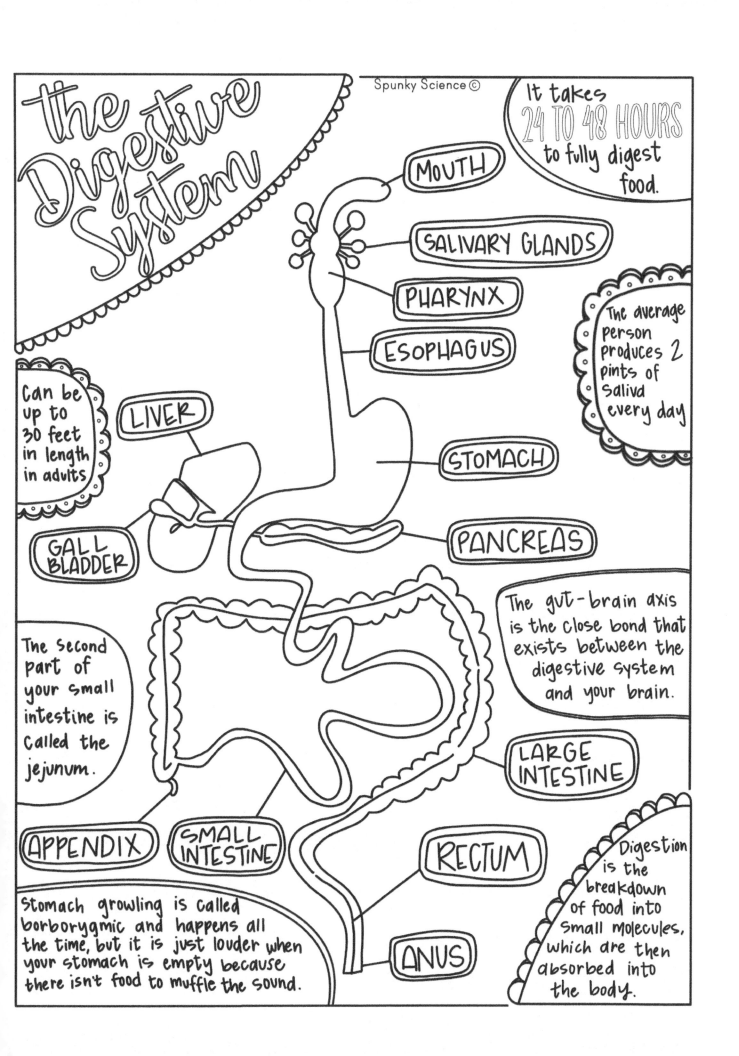

the Digestive System

Spunky Science ©

It takes **24 TO 48 HOURS** to fully digest food.

MOUTH

SALIVARY GLANDS

PHARYNX

ESOPHAGUS

The average person produces 2 pints of saliva every day

LIVER

Can be up to 30 feet in length in adults

STOMACH

GALL BLADDER

PANCREAS

The gut-brain axis is the close bond that exists between the digestive system and your brain.

The second part of your small intestine is called the jejunum.

LARGE INTESTINE

APPENDIX

SMALL INTESTINE

RECTUM

Digestion is the breakdown of food into small molecules, which are then absorbed into the body.

Stomach growling is called borborygmic and happens all the time, but it is just louder when your stomach is empty because there isn't food to muffle the sound.

ANUS

INTEGUMENTARY SYSTEM

THE INTEGUMENTARY SYSTEM ACTS TO PROTECT THE BODY FROM *damage*, WATER LOSS AND *infection*. IT INCLUDES SKIN, HOOVES, HAIR, AND FEATHERS.

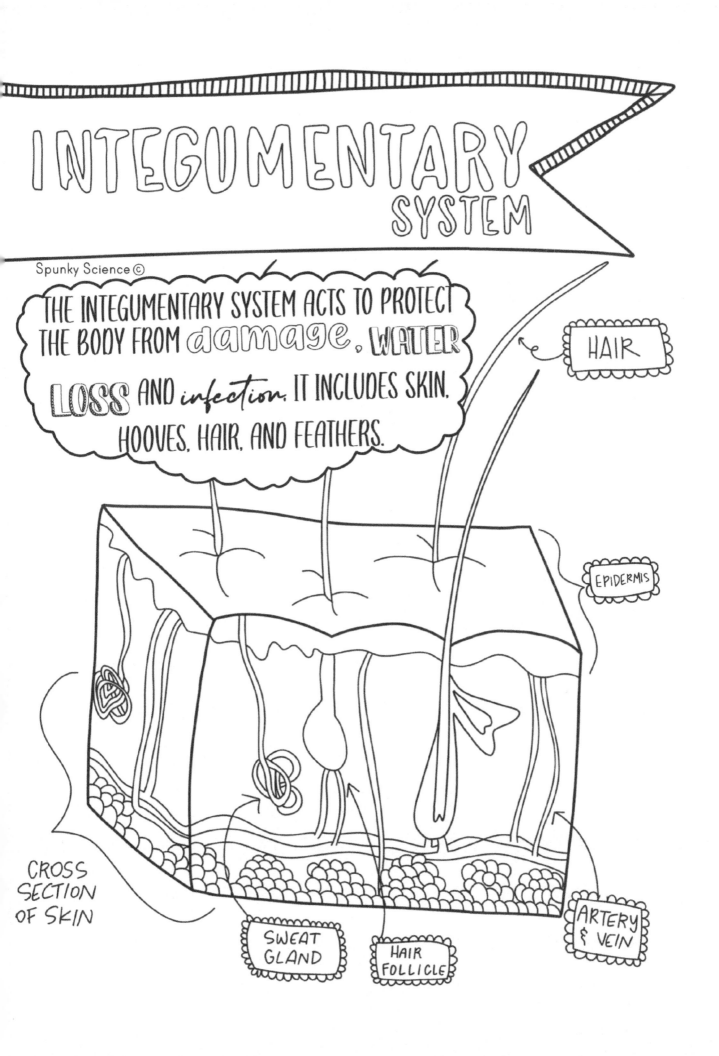

HAIR

EPIDERMIS

CROSS SECTION OF SKIN

SWEAT GLAND

HAIR FOLLICLE

ARTERY & VEIN

EXCRETORY SYSTEM

The systems that excrete wastes from the body. For example, the system of organs that regulates the amount of water in the body and filters and eliminates from the blood the wastes produced by metabolism. The principal organs of the excretory system are the kidneys, ureters, urethra, and urinary bladder.

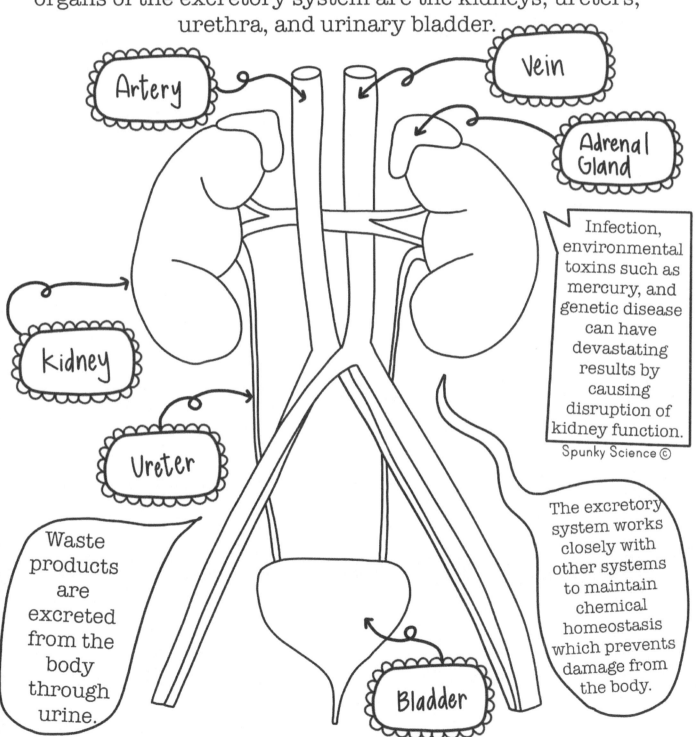

Artery

Vein

Adrenal Gland

Kidney

Ureter

Infection, environmental toxins such as mercury, and genetic disease can have devastating results by causing disruption of kidney function.

Spunky Science ©

Waste products are excreted from the body through urine.

The excretory system works closely with other systems to maintain chemical homeostasis which prevents damage from the body.

Bladder

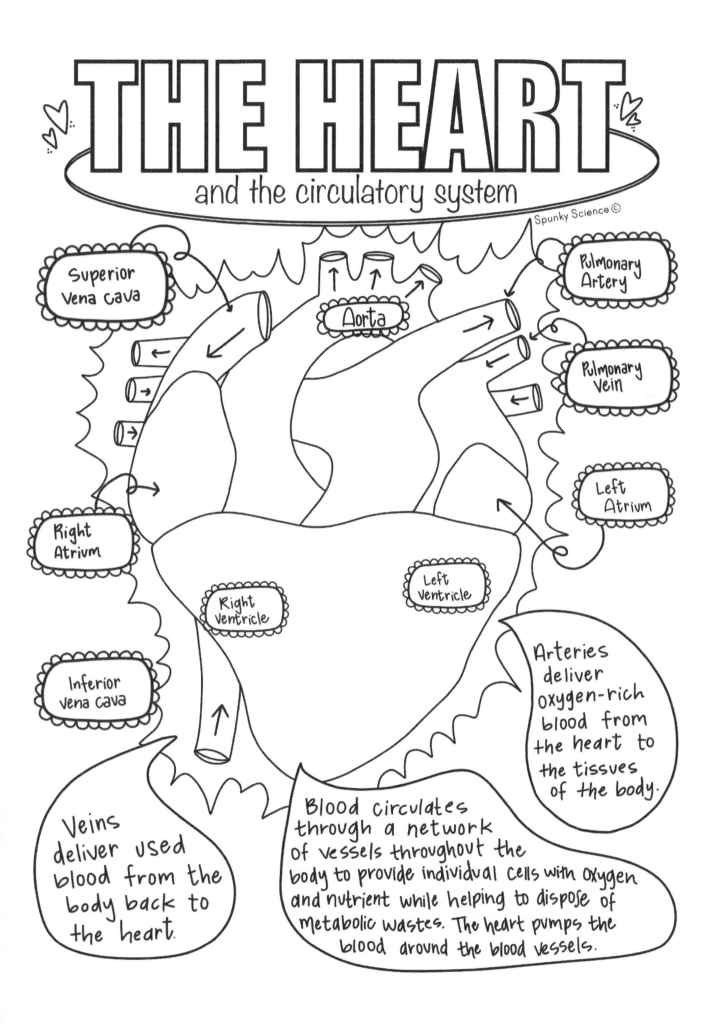

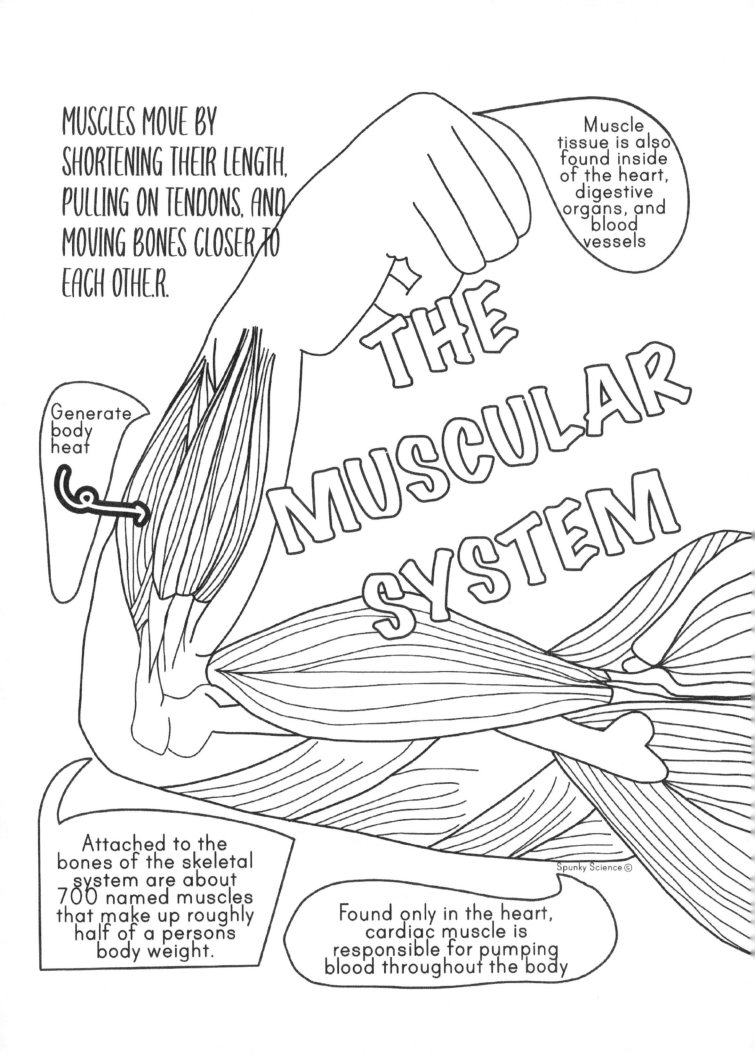

Thyroid

THE THYROID GLAND LOOKS LIKE A *butterfly* AND IS LOCATED AT THE FRONT OF THE NECK BELOW THE ADAM'S APPLE.

ALMOST EVERY **CELL** IN YOUR BODY NEEDS THYROID HORMONES

SOMETIMES THE THYROID IS *hyperactive* AND PRODUCES TOO MUCH THYROID HORMONE.

TO WORK PROPERLY, THE THYROID NEEDS IODINE.

AMONG MANY OTHER THINGS, THYROID HORMONES HELP **control** HOW FAST YOU BURN CALORIES, HOW FAST YOUR HEART BEATS, AND YOUR BODY TEMPERATURE.

Anxiety AND *insomnia* CAN BE SIGNS OF AN OVERACTIVE THYROID

THE THYROID IS UNDER THE *control* OF A PEANUT SHAPED GLAD IN THE BRAIN CALLED THE **PITUITARY** GLAND.

Spunky Science ©

FETAL DEVELOPMENT and the REPRODUCTIVE SYSTEM

2-Cell Stage

Fertilized egg

Fetus– 20 weeks

4-Cell Stage

8-cell stage

Fetus–16 weeks

16–Cell Stage

Fetus–4 weeks

Fetus–10 weeks

Blastocyst

WHY WE STUDY
TAXONOMY

IT IS IMPORTANT TO CLASSIFY LIVING THINGS BECAUSE...

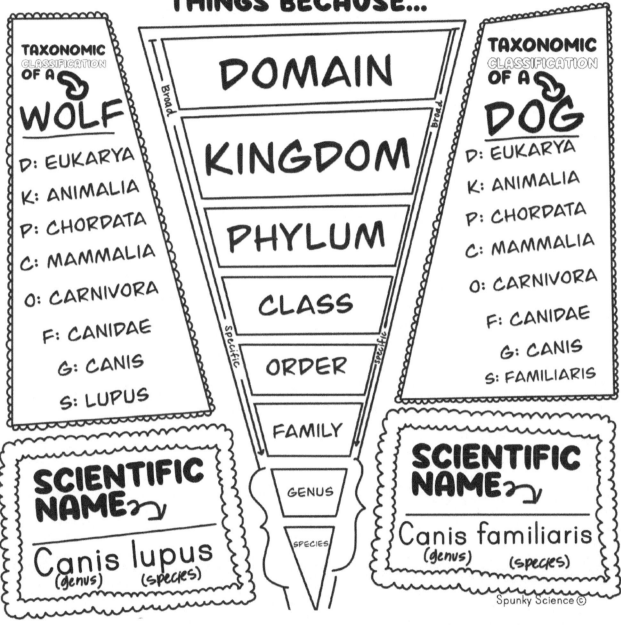

TAXONOMIC CLASSIFICATION OF A

WOLF

D: EUKARYA
K: ANIMALIA
P: CHORDATA
C: MAMMALIA
O: CARNIVORA
F: CANIDAE
G: CANIS
S: LUPUS

TAXONOMIC CLASSIFICATION OF A

DOG

D: EUKARYA
K: ANIMALIA
P: CHORDATA
C: MAMMALIA
O: CARNIVORA
F: CANIDAE
G: CANIS
S: FAMILIARIS

Broad.   broad.

DOMAIN

KINGDOM

PHYLUM

CLASS

specific   specific

ORDER

FAMILY

GENUS

SPECIES

SCIENTIFIC NAME

Canis lupus
(genus)   (species)

SCIENTIFIC NAME

Canis familiaris
(genus)   (species)

Spunky Science ©

WITHOUT IT WE WOULD NOT UNDERSTAND HOW VARIOUS ORGANISMS ARE SIMILAR AND DIFFERENT

KINGDOM PLANTAE

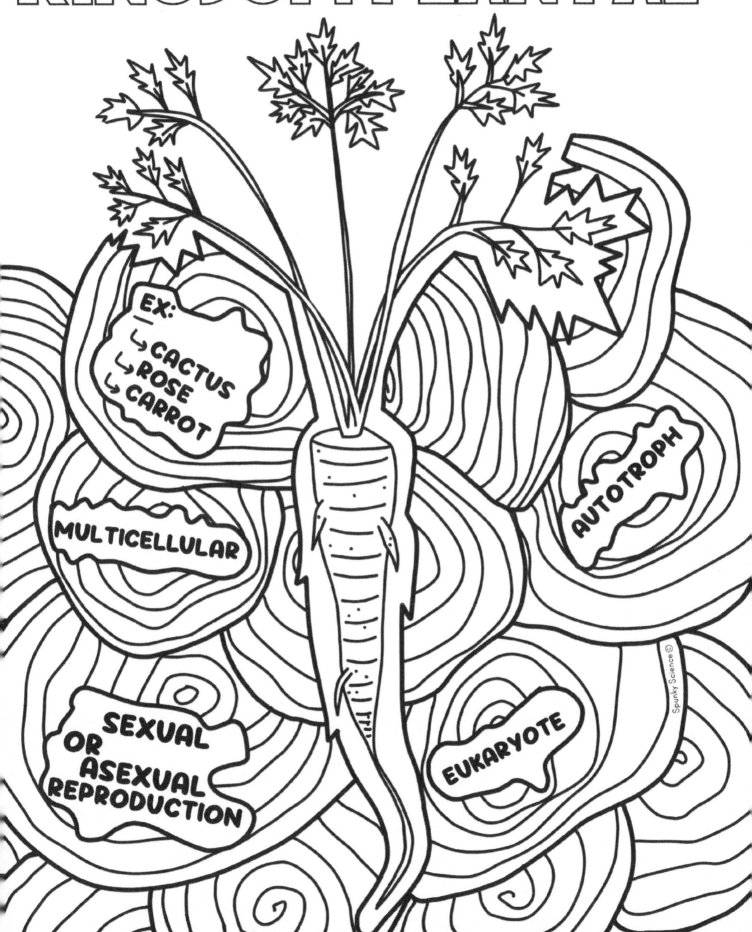

EX:
↳ CACTUS
↳ ROSE
↳ CARROT

MULTICELLULAR

AUTOTROPH

SEXUAL OR ASEXUAL REPRODUCTION

EUKARYOTE

Spunky Science ©

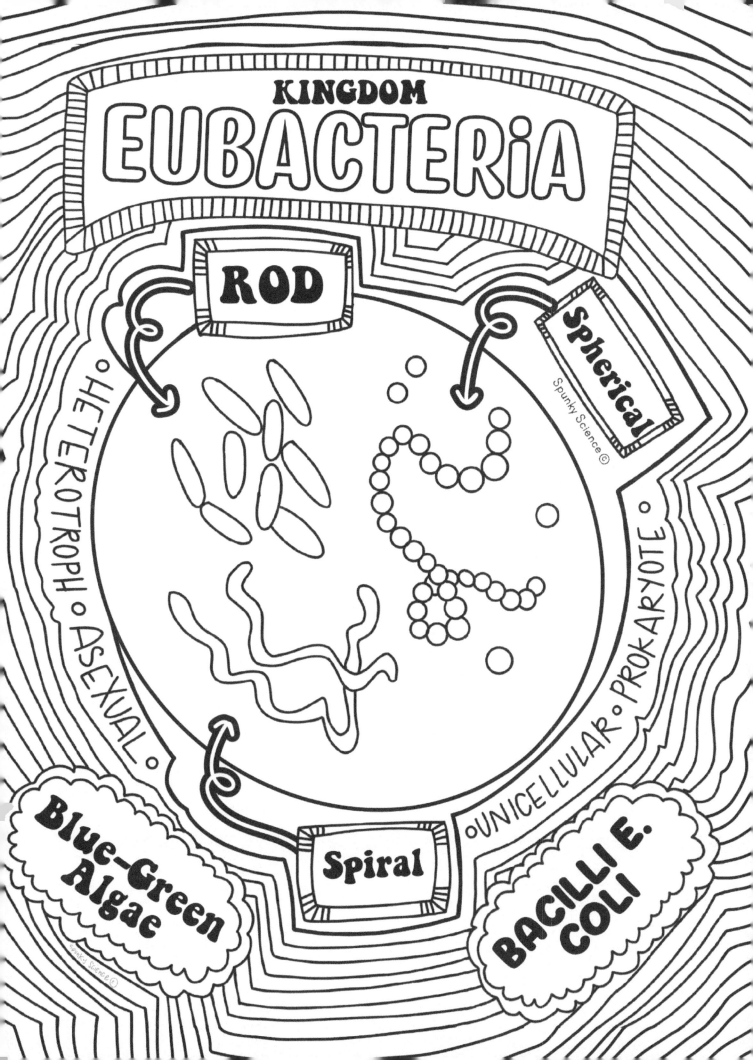

MUTUALISM AND THE RAINBOW PARROTFISH

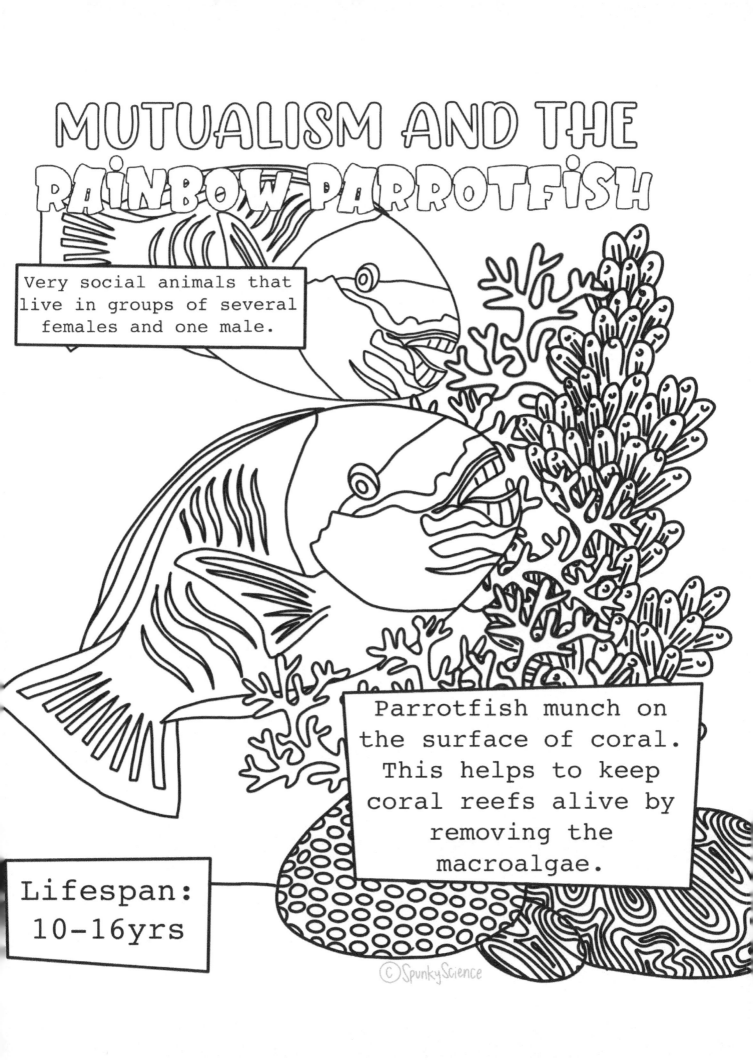

Very social animals that live in groups of several females and one male.

Parrotfish munch on the surface of coral. This helps to keep coral reefs alive by removing the macroalgae.

Lifespan: 10-16yrs

COMMENSALISM AND THE CLOWN FISH

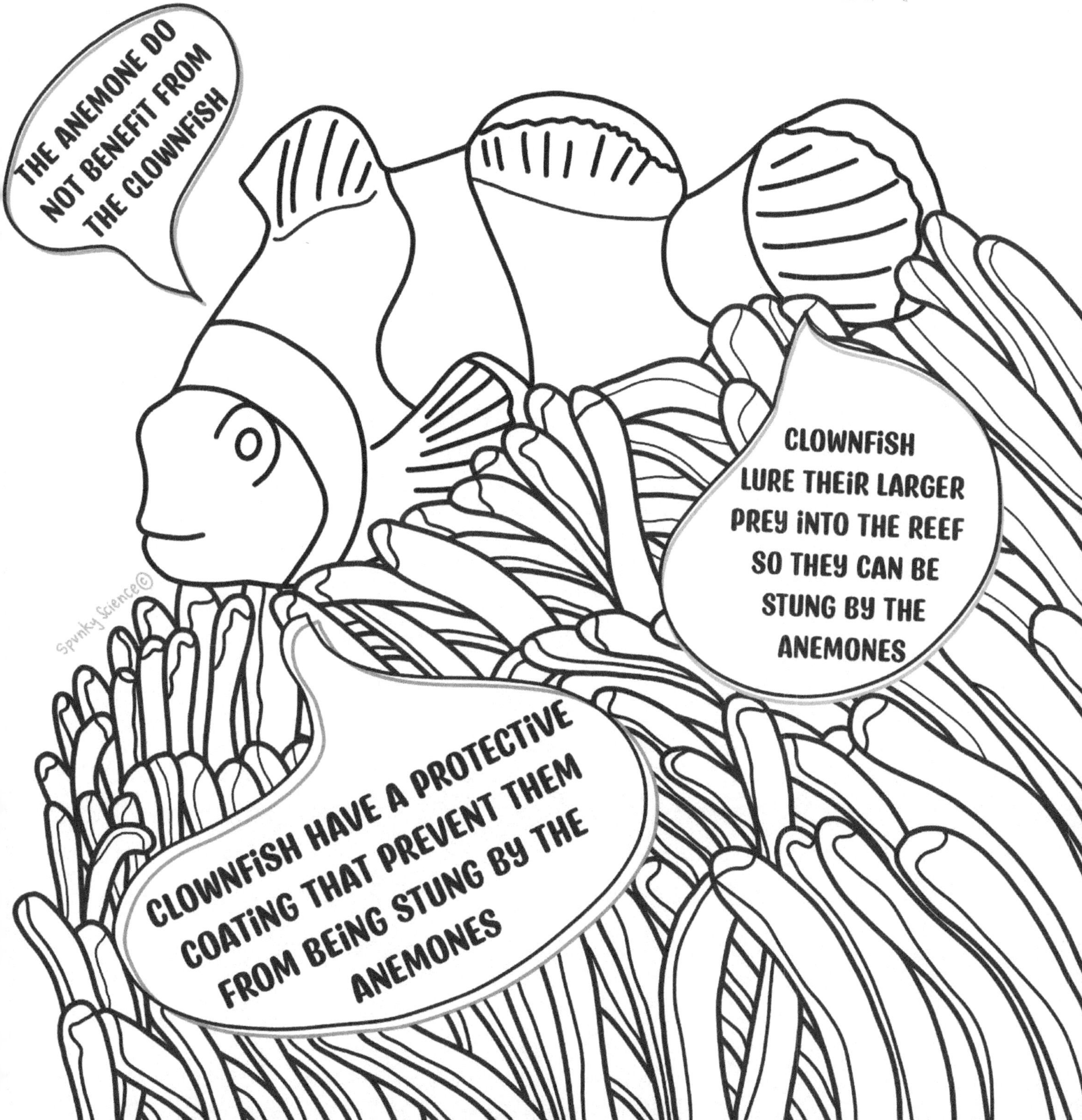

ORGANISM INTERACTIONS

PREDATORY

The pursuit, capture, and killing of animals for food.

Puffer fish take in water to look big in hopes of scaring away predators.

COMPETITIVE

Penguins and winged birds

Big fishes

Two or more organisms compete for limited resources, such as nutrients, living space, or light.

SYMBIOTIC

A relationship in which two species live closely together; can be symbiotic, commensalism, or mutualistic relationships.

ROUNDWORM IN YOUR EYE

Roundworm

Spunky Science ©

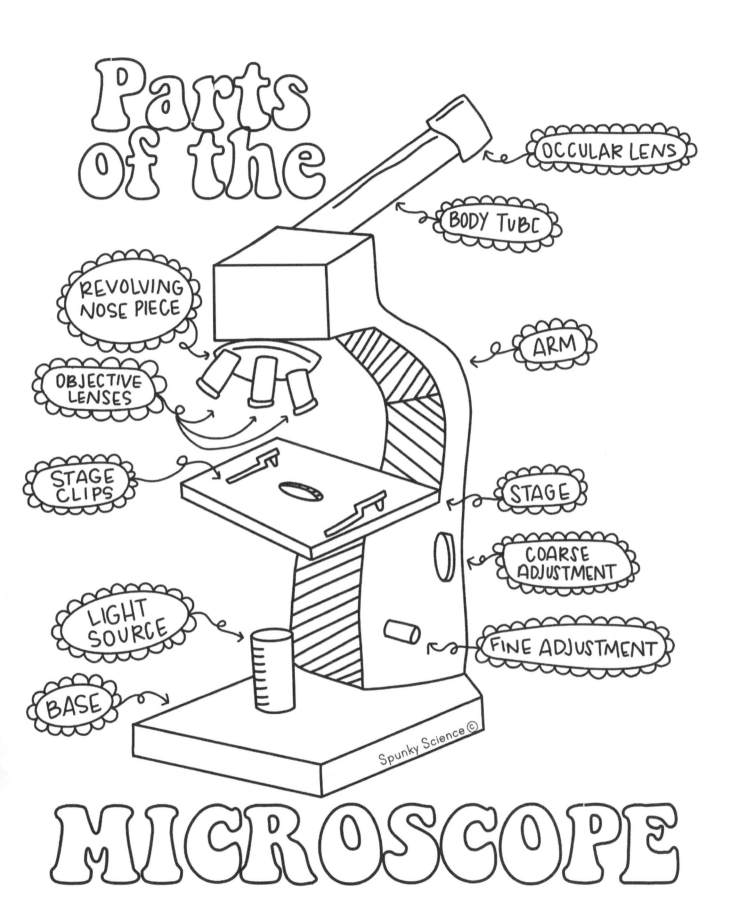

CELL THEORY

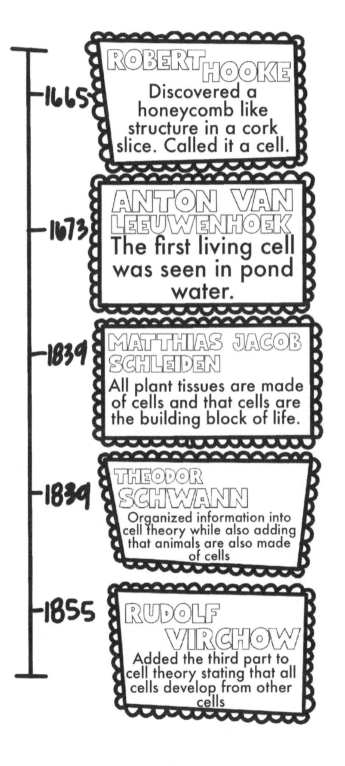

1665 — ROBERT HOOKE
Discovered a honeycomb like structure in a cork slice. Called it a cell.

1673 — ANTON VAN LEEUWENHOEK
The first living cell was seen in pond water.

1839 — MATTHIAS JACOB SCHLEIDEN
All plant tissues are made of cells and that cells are the building block of life.

1839 — THEODOR SCHWANN
Organized information into cell theory while also adding that animals are also made of cells

1855 — RUDOLF VIRCHOW
Added the third part to cell theory stating that all cells develop from other cells

3 Principles

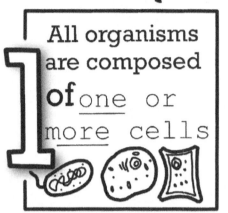

1 All organisms are composed of <u>one or more cells</u>

2 The cell is the basic unit of LIFE

3 All cells come from pre-existing cells

©Spunky Science

EUKARYOTES

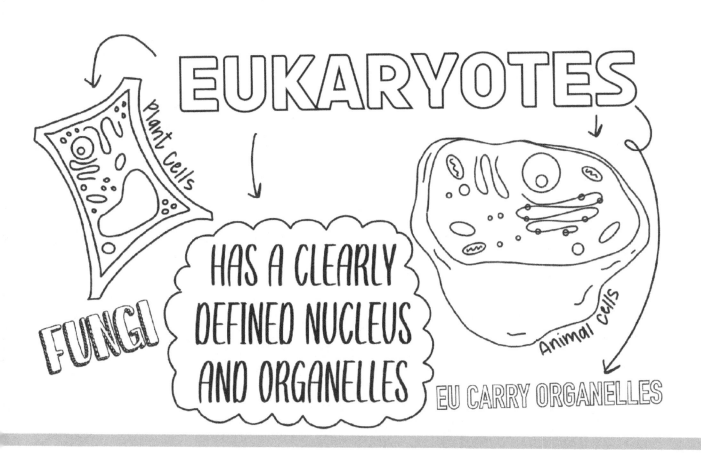

Plant cells

FUNGI

HAS A CLEARLY DEFINED NUCLEUS AND ORGANELLES

Animal cells

EU CARRY ORGANELLES

PROKARYOTES

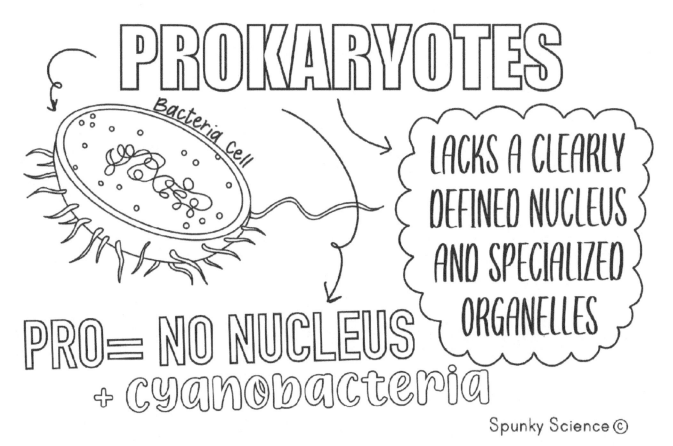

Bacteria Cell

LACKS A CLEARLY DEFINED NUCLEUS AND SPECIALIZED ORGANELLES

PRO= NO NUCLEUS + cyanobacteria

Spunky Science ©

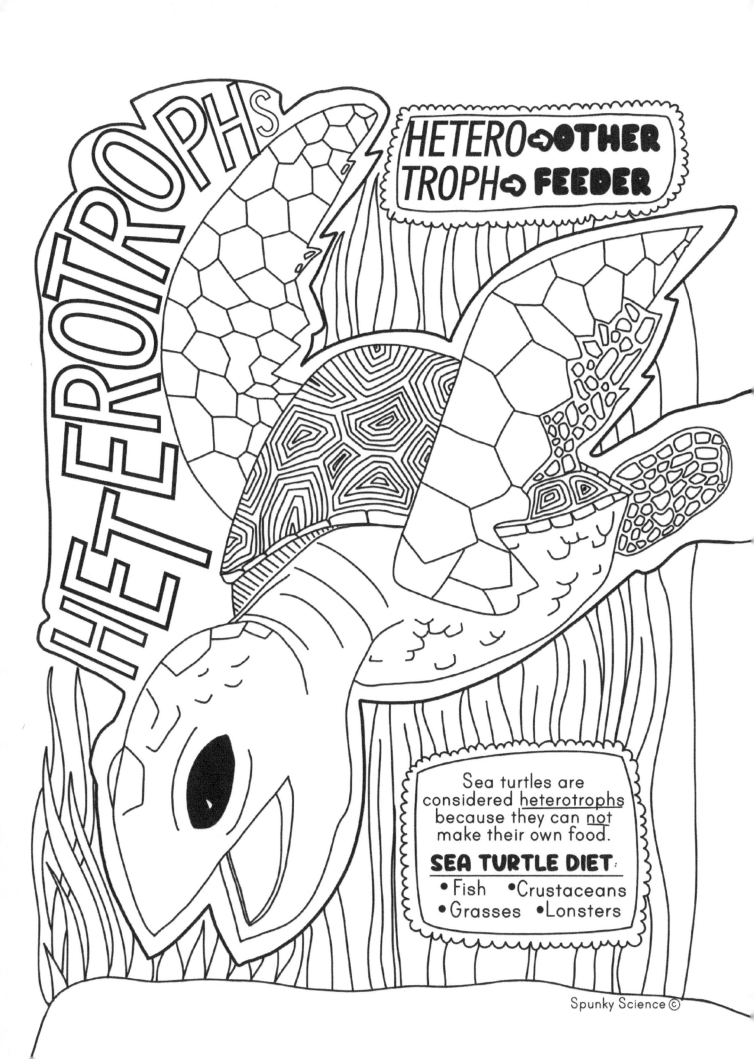

HETEROTROPHS

HETERO → OTHER
TROPH → FEEDER

Sea turtles are considered <u>heterotrophs</u> because they can <u>not</u> make their own food.

SEA TURTLE DIET:
- Fish
- Crustaceans
- Grasses
- Lonsters

Spunky Science ©

AUTOTROPH

AUTO SELF

TROPH FEEDER

Autotrophs take in energy from the sun in order to make their own food by **PHOTOSYNTHESIS.**

WHAT'S A VIRUS?

A VIRUS IS AN INFECTIVE AGENT THAT TYPICALLY CONSISTS OF A NUCLEIC ACID MOLECULE IN A PROTEIN COAT, IS TOO SMALLL TO BE SEEN BY LIGHT MICROSCOPY, AND IS ABLE TO MULTIPLY ONLY WITHIN THE LIVING CELLS OF A HOST.

Spunky Science ©

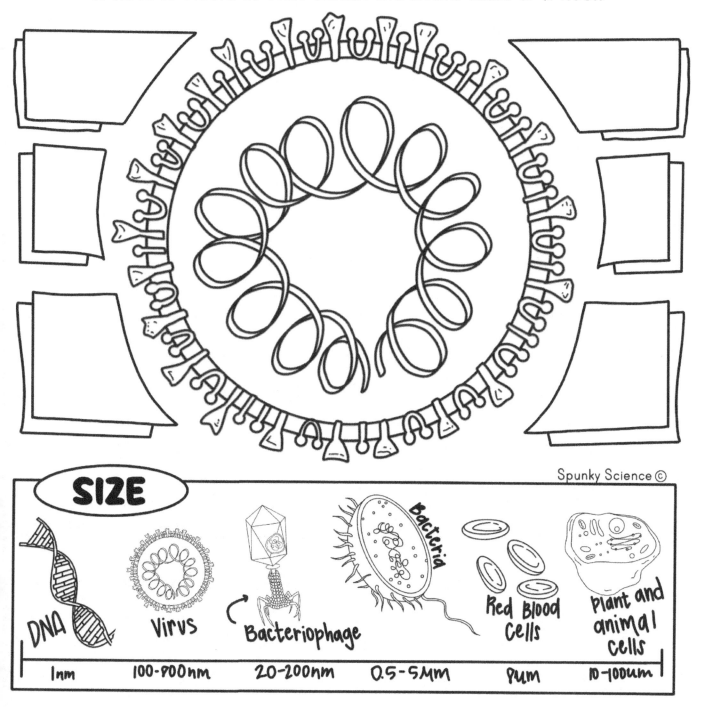

SIZE

DNA	Virus	Bacteriophage	Bacteria	Red Blood Cells	Plant and animal cells
1nm	100-900nm	20-200nm	0.5-5μm	8μm	10-100μm

Stages of MITOSIS

1 Interphase

2 Prophase

Metaphase 3

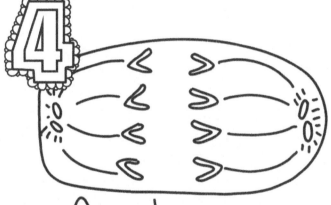

4 Anaphase

Telophase 5

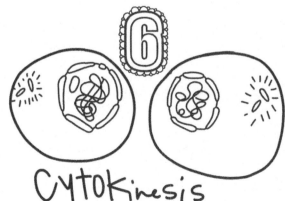

6 Cytokinesis

Spunky Science ©

BACTERIA CELL

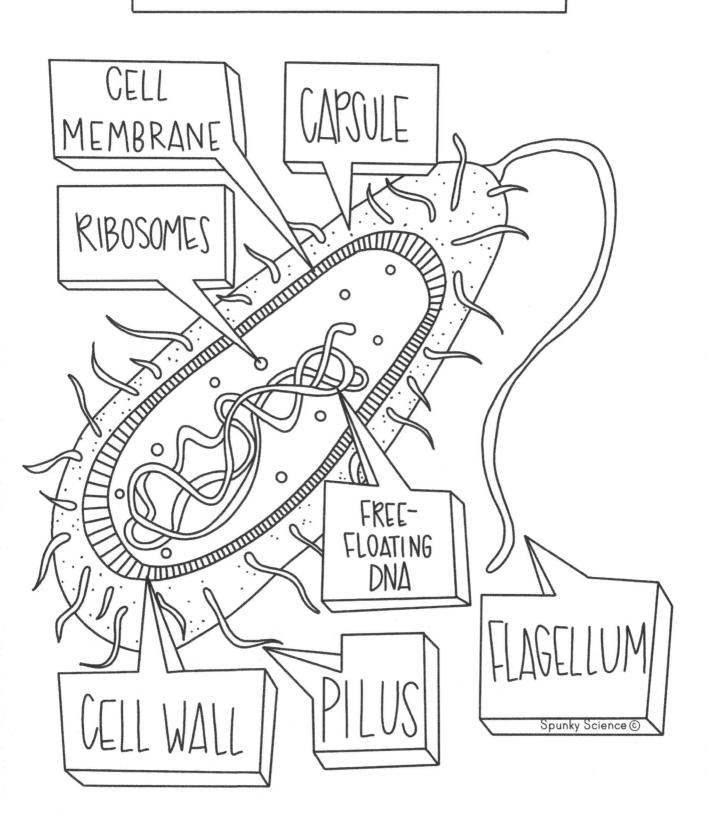

CELL MEMBRANE

CAPSULE

RIBOSOMES

FREE-FLOATING DNA

CELL WALL

PILUS

FLAGELLUM

Spunky Science ©

PLANTCELL

Plant cells are the basic unit of life in the kingdom plantae. They are eukaryotic cells, which have a true nucleus along with specialized structures called organelles that carry out different functions.

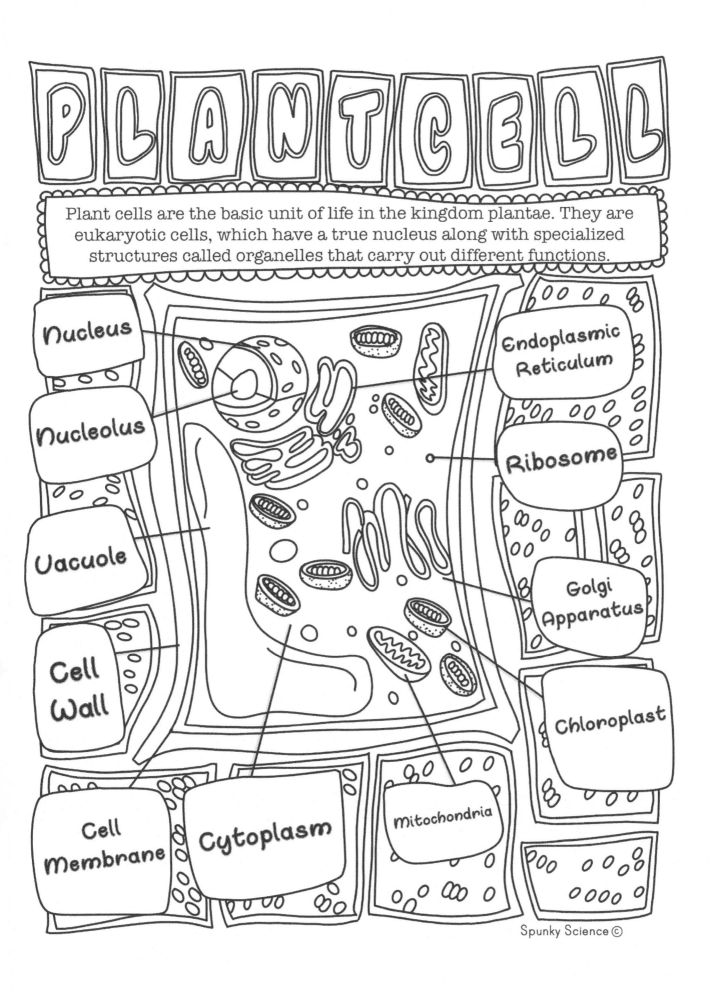

Nucleus

Nucleolus

Vacuole

Cell Wall

Endoplasmic Reticulum

Ribosome

Golgi Apparatus

Chloroplast

Cell Membrane

Cytoplasm

mitochondria

Spunky Science ©

3 TYPES OF MUSCLE TISSUES

skeletal Tissue

- Voluntary control of the nervous system
- Attached to bones by tendons

smooth Tissue

stomach

- Located in the walls of hollow visceral organs except the heart
- Involuntary control of the nervous system

cardiac Tissue

- In the walls of the heart
- Striated
- Under involuntary control

Made in the USA
Las Vegas, NV
01 February 2024

85185263R00063